Introduction

This manual gives the experimenter a basic understanding on how diodes can be utilized, starting with basic power supply circuits and working up to more complex circuits such as frequency multipliers and modulators. On first glance, the diode looks like a very simple device but has many useful characteristics that can be taken advantage of outside of the basic rectifier. This manual doesn't try to be a complete source of information about a very complex subject but merely scratches the surface and gives some starting points where one can work up to more complex and sophisticated designs.

Wherever possible, the technical jargon has been kept to a minimum and the schematic diagram is explained in a straight forward manner. With most of the schematics, oscilloscope screens are given to further explain what is happening in the particular circuit.

Brief history

A diode is an electronic device that allows electric current to flow in one direction with low resistance but offers high resistance to the flow of electric current in the opposite direction. This effect has proven to be useful in telecommunications and power electronics. The solid state (semiconductor) and vacuum tube type diode were gradually developed as parallel technologies from the late 1800's onward. In 1880 while trying to improve his light bulb, Thomas Edison put another wire connected to an electrode into the glass bulb and put a positive charge on it. He noticed that DC current would flow through the vacuum from the filament (cathode) to the anode (electrode) but not the other direction. Not finding a use for it at the time, he patented it in 1884. It became known as the "Edison Effect"

Later, English electrical engineer and physicist, John Ambrose Fleming realized the phenomenon would make a good detector for radio signals and he patented the "Fleming Valve" in Britain in 1904 and the United States in 1905.

The main problem with the Fleming valve was that it needed a substantial DC voltage (DC Bias) across it before it could be suitable for the detection of radio signals. Putting sufficient bias across the tube allowed miniscule radio signals (that might be in the order of microvolts) to pass through the tube unimpeded in one direction, but be blocked in the other direction.

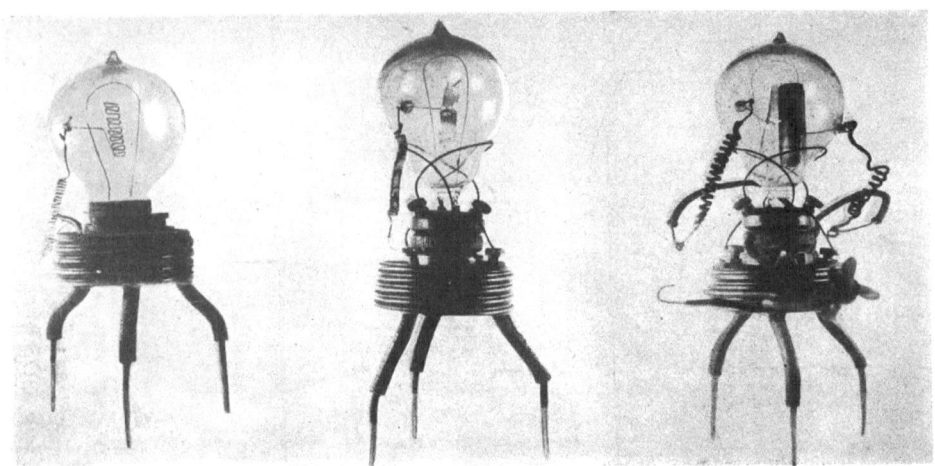

Prototype Fleming Valves c. 1904

German Physicist Karl Ferdinand Braun patented a crystal rectifier in 1899 which was later called the "Cat's Whisker Detector" used in early crystal receiving sets. This consisted of a Galena (Lead Sulfide) crystal in a metal holder where the holder was connected to one terminal and the fine wire or the cat's whisker was connected to another terminal. The wire was moved over the crystal until a signal was received with the greatest sensitivity. Early American radio researcher Greenleaf Whittier Pickard received a patent in 1906 for the silicon diode. In this invention, a fine wire was permanently connected to a piece of silicon eliminating the need to find a "sensitive spot".

Greenleaf Whittier Pickard's silicon diode c.1906

The one advantage that the Galena crystal had was its low barrier potential which is below .2 volts. A silicon diode has a barrier potential of .6 volts so the Galena crystal can detect much lower voltage and is therefore more sensitive in a radio receiver that has no amplification. In the early days of radio, many different types of materials were experimented with and some were found to have better sensitivity than others.

The Fleming Valve became the choice of rectifier in commercial and consumer radio receivers for the next 50 years for both signal detection and power rectification. By the 1960's when the transistor became the amplifying device of choice, the encapsulated Germanium point contact diode was used for detection in most transistor radios. Encapsulated silicon diodes were developed but used mainly in power supplies. These completely replaced the inferior Selenium power rectifiers that were used in industrial equipment and televisions from the 1930's to the 1950's. Any old time electronics technician could tell you about the unique smell of a burnt out selenium rectifier!

Two types other of rectifiers that saw use in industrial applications were the copper oxide and mercury arc types. The copper oxide type found its use in battery chargers and the mercury arc type was used in applications such as DC power supplies for tramcars and electric buses.

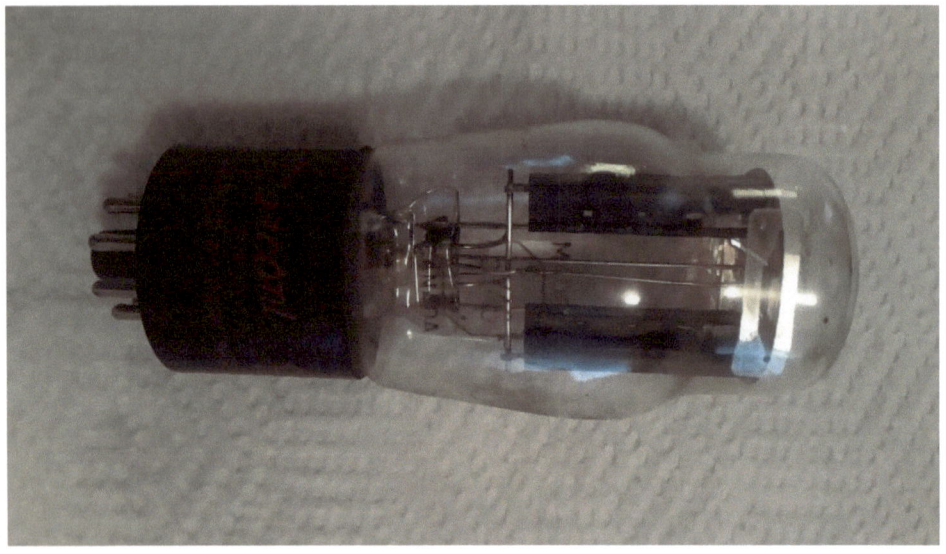

Canadian made 5Y4G Full Wave Rectifier from 1938. This tube has a filament which acts as a common cathode to two anodes. Was most commonly found in larger radios and televisions up to the 1960's.

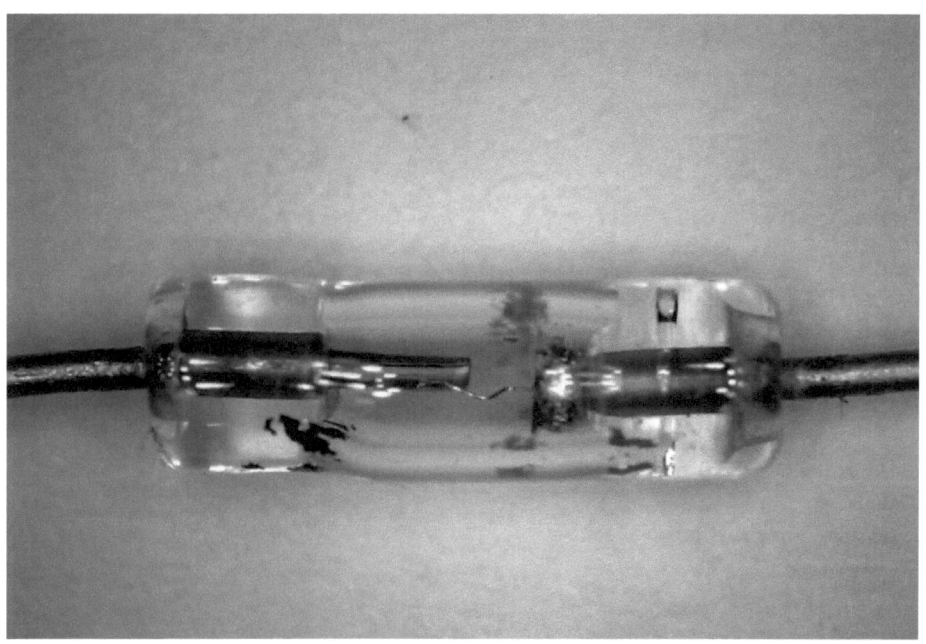

1N34A glass encapsulated germanium signal diode, magnified 200 times. Note the thin wire or "cat's whisker" attached to a block of germanium to form a PN junction. The barrier potential of this diode is .25 to .30 volts.

The germanium point contact diode was refined during WW2 for the need to have a sensitive detector for radar sets. The low capacitance and fast recovery of the germanium point contact diode made it ideally suited for microwave work.

Since the Germanium point diode only has a mechanical connection between the cat's whisker and Germanium, it's prone to produce noise due to mechanical vibration. For demanding military applications, it's been superseded by Schottky diodes designed for microwave frequencies.

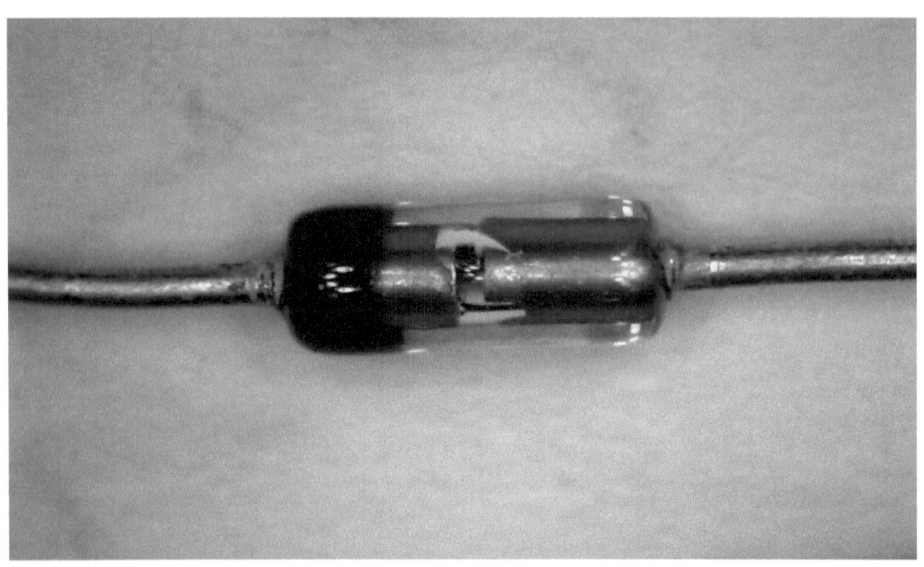

1N914/1N4148 glass encapsulated silicon switching diode, magnified 200 times. Note the different arrangement of PN junction from the germanium signal diode. The barrier potential of this diode is .50 to .60 volts.

Since WW2, many different types of specialty diodes have been developed in which the only property that justifies their being called "diodes" is the fact that they are active devices and have two electrodes. Some of these types include Avalanche diodes, Light-Emitting diodes, photodiodes, PIN diodes, Photodiodes, Schottky diodes and Zener diodes. In this book we will explore uses of the 1N4xx series, the 1N914/1N4148 and some common Zener diodes.

Glossary of diode terminology

It is useful to have some basic theory to help you understand the nature of diodes and why they behave the way they do. Following are some terms you will see on a diode data sheet from the manufacturer and their definitions.

Anode
The terminal that is denoted by an arrow in the diode schematic it's normally connected to the positive side of a circuit to allow for current flow though the device.

Bias
This is the voltage needed to turn on the device, .2 volts in the case of a Schottky diode, .3V in the case of a Germanium and .7V in the case of a Silicon diode.

Peak Inverse Voltage or Reverse Breakdown Voltage
(PIV) or Maximum Reverse Voltage VR (max) is the maximum allowable reverse voltage that can be applied across the device without reverse breakdown and damage occurring to the device. The 1N400x series of diodes have ratings of 50V in the case of 1N4001 to 1000V in the case of 1N4007.

Cathode
The terminal that is negative relative to the anode. The cathode is denoted by a vertical line touching the anode arrow.

Forward voltage drop
This is the voltage measured across a conducting forward biased diode. The voltage remains relatively constant regardless of current flow. The voltage is approximately .7 volts in the case of a silicon diode and .3 volts in the case of a germanium diode.

Recovery time
When a diode is switching from its conducting state to its blocking state, a diode or rectifier must get rid of its stored charge before it can block reverse current again. The discharge takes a certain measurable amount of time and in the case of the 1N4148 it is from 4 to 8 nanoseconds. Switching diodes such as the 1N4148 are designed to have as quick a recovery time as possible.

Total Power Dissipation
PD (max) is the maximum total power dissipation of the diode when it is forward biased. In the case of the 1N4148 the maximum device dissipation is 500 mW.

Maximum Forward Current
This is the maximum permissible current that the device should conduct when it is forward biased. There is a small resistance across the PN junction and according to Ohm's law there will be some power dissipated across the junction and therefore waste heat. As long as the rating is not exceeded there should be no problem with overheating.

Diodes used in power supply Circuits

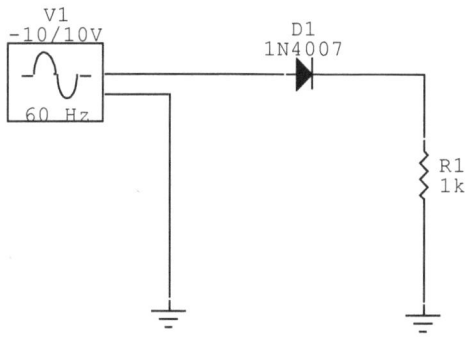

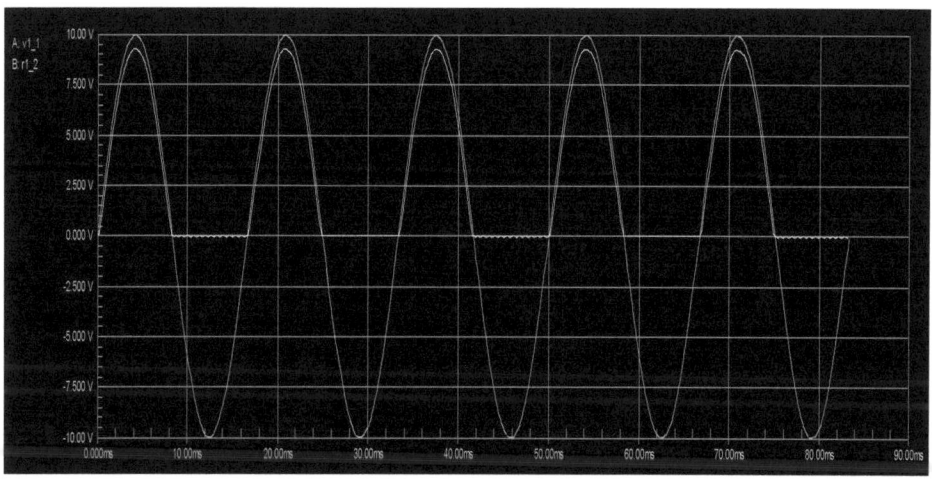

Half Wave Rectifier

The half wave rectifier circuit shown above is the simplest form of rectifier. The circuit consists of a simple diode and capacitor. As the AC voltage is applied to the input of the diode the current is allowed to flow in one direction and not in the other. This results in a pulsating DC current at the output of the diode. A large value electrolytic capacitor is put across the output of the diode and ground to smooth out the AC component. This is not a good circuit to use where the AC component is going to affect the performance of the circuit being powered. This type of rectifier is best used in applications such as arc welders.

The image shows the input waveform (green) and the output waveform (yellow). The output waveform shows some noticeable AC component (ripple) which is the same as the AC input frequency, 60 Hz.

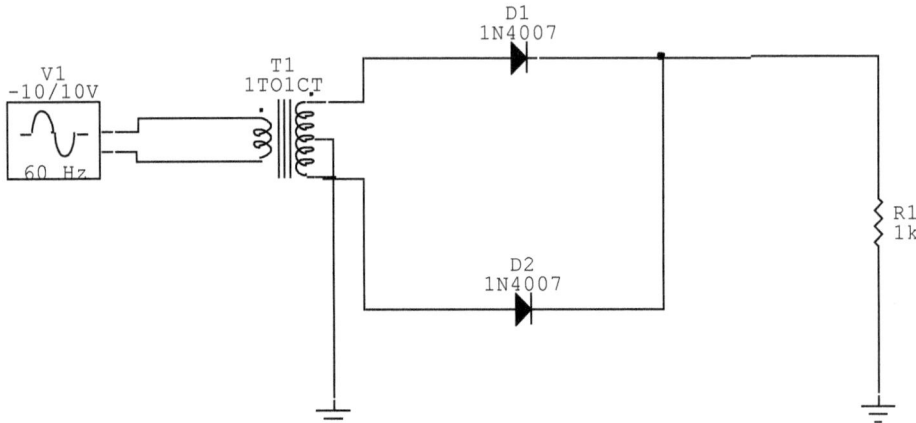

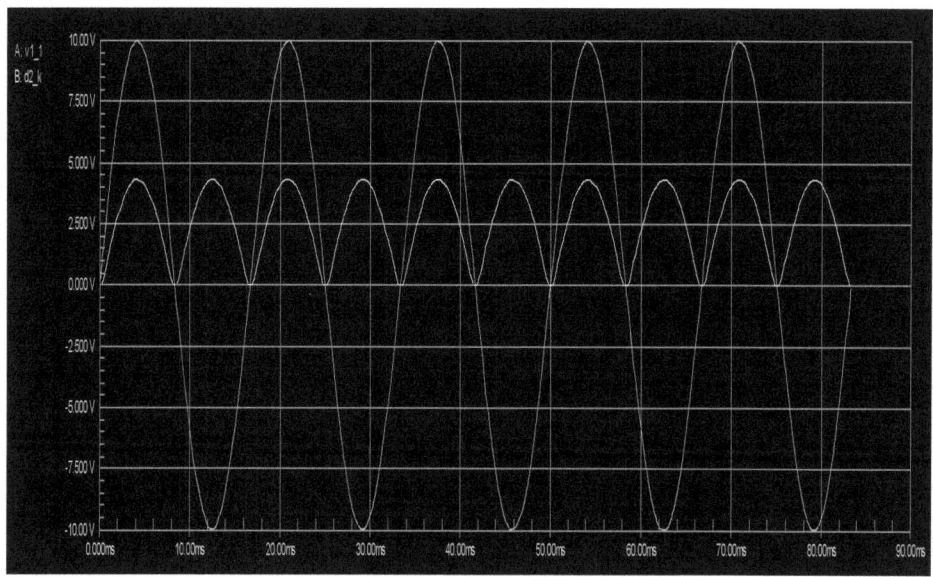

Full Wave Rectifier

The full wave rectifier circuit shown on the previous page gives much better performance than the half wave circuit. Notice that the AC component (yellow) is at twice the input frequency (green). Assuming that the components are 100 % efficient, this type of rectifier theoretically gives 100 % conversion of the AC current to DC. The half wave circuit can never be over 50 % efficient because half the AC waveform is cut off. In the real world there are losses in the transformer and diodes due to various factors. A more realistic efficiency for half wave circuit would be somewhere in the 40-45 % region and the full wave circuit efficiency would be in the 80-90 % region. A full wave rectifying circuit where the expensive transformer is dispensed with and replaced with two much cheaper diodes, is called a bridge full wave rectifier. This type of circuit is usually fabricated in a package where four matched diodes are already wired together with just two AC input terminals and two DC output terminals.

Full Wave Bridge Rectifier

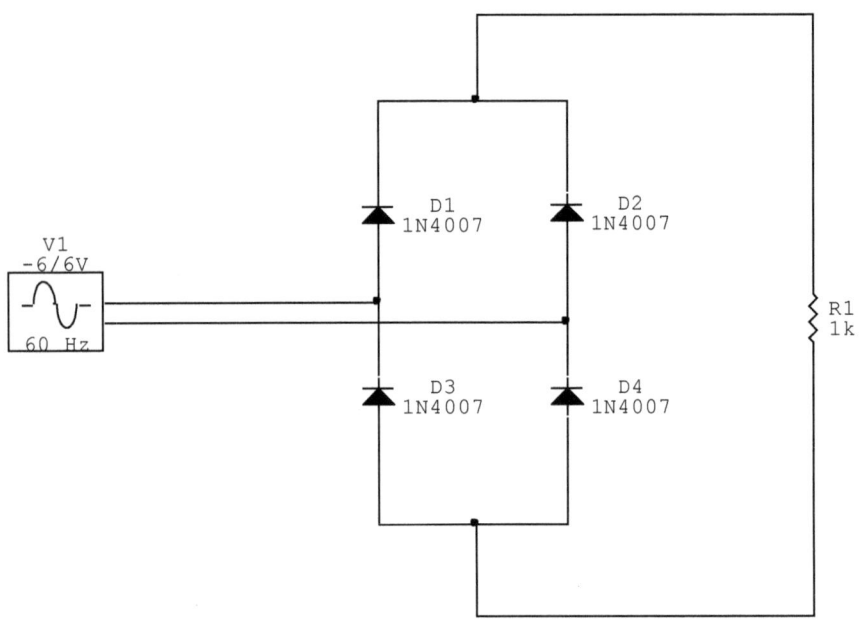

Gives the same output waveform
as the previous circuit

Voltage Multiplication Circuits

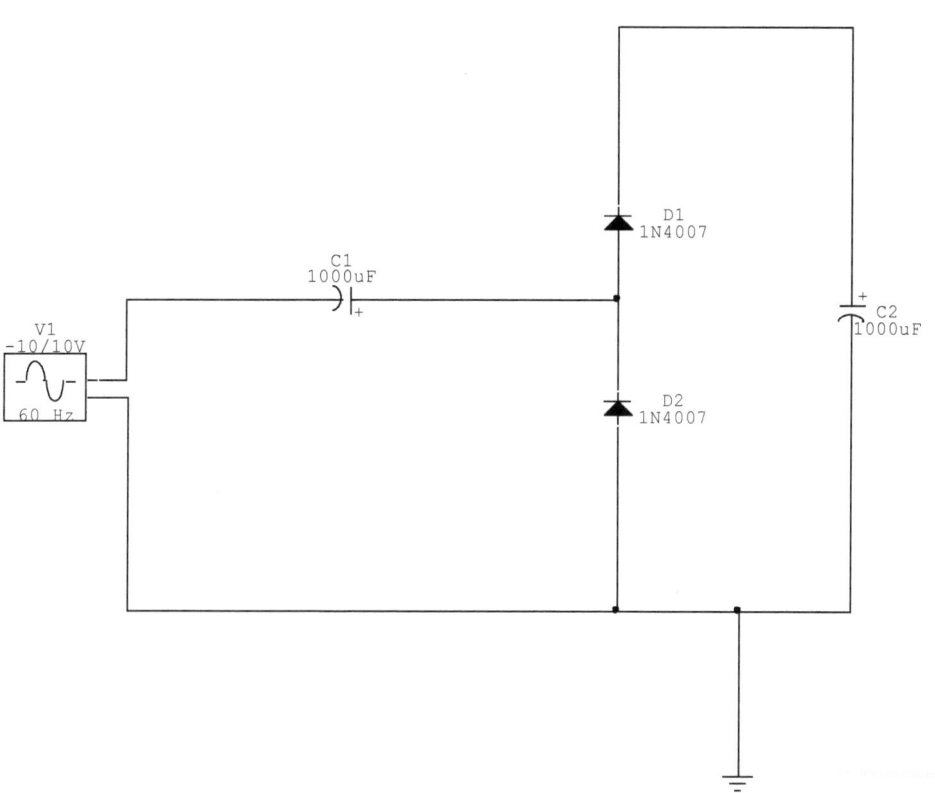

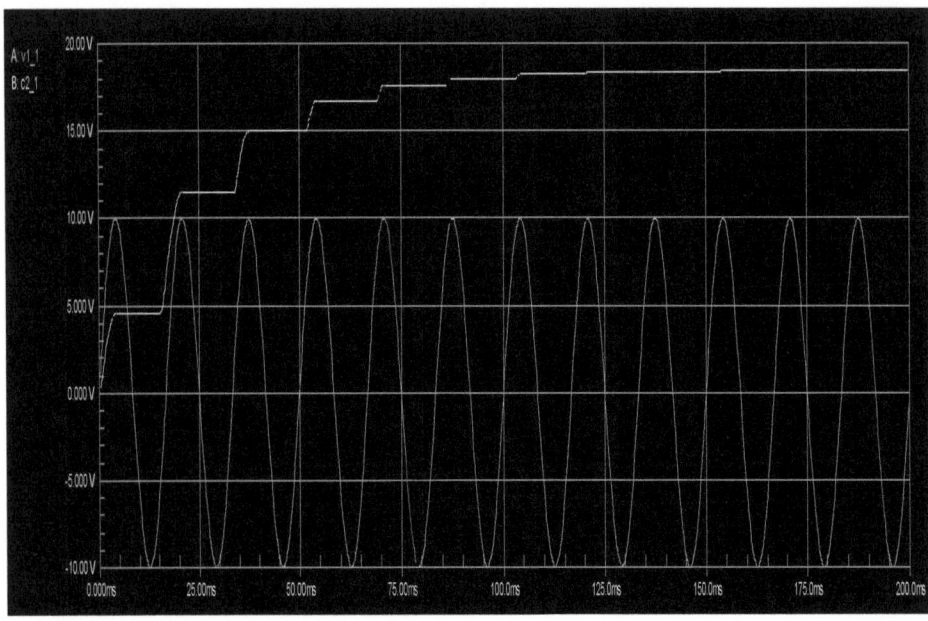

Half Wave Voltage Doubler

In the circuit on the previous page, the (no load) DC voltage across the capacitor C1 is 2.8 x input RMS voltage less the diode drop. When source goes negative, C2 charges through D2. When source goes positive, C2 is in series with the source, D2 blocks C2 discharge, C1 gets charged through D1.

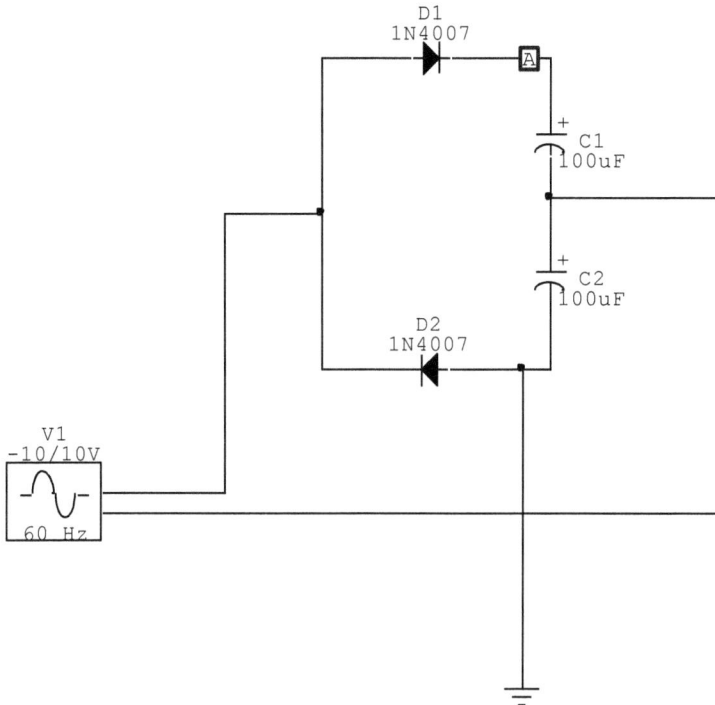

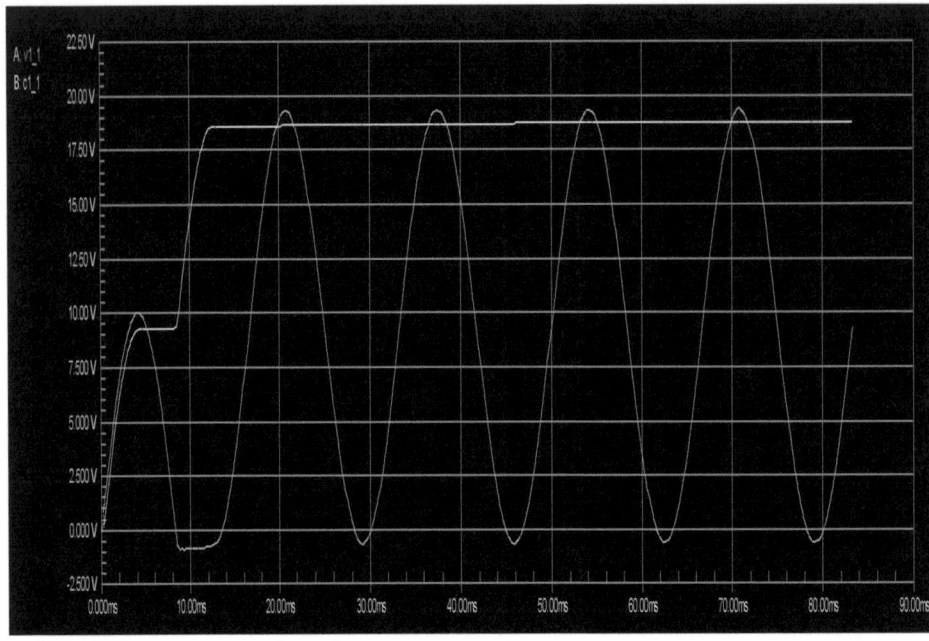

Full Wave Voltage Doubler

In the circuit on the previous page, diode D2 charges capacitor C2 on the negative half cycle of input. Diode D1 charges capacitor C1 on the positive half cycle. The total voltage taken across C1 and C2 is 2.8 x the RMS voltage less the diode voltage drops.

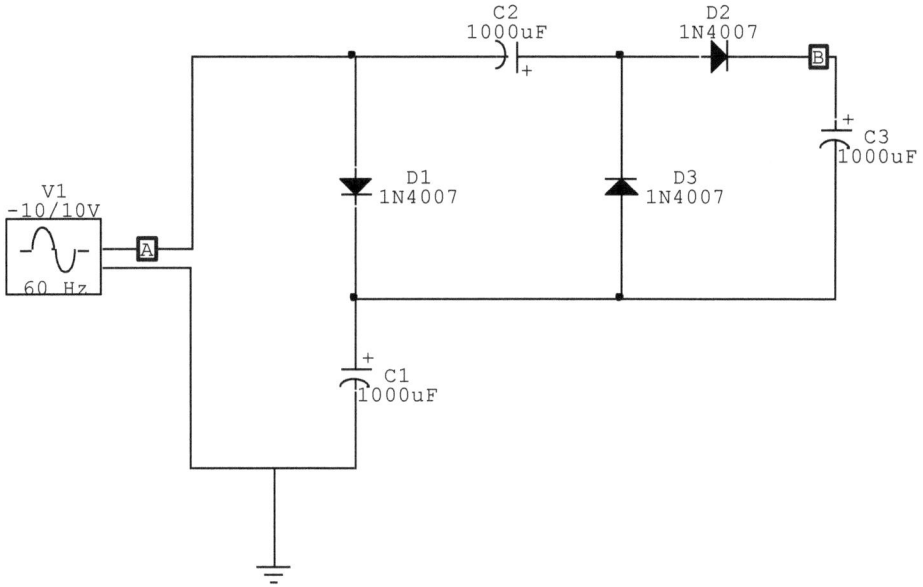

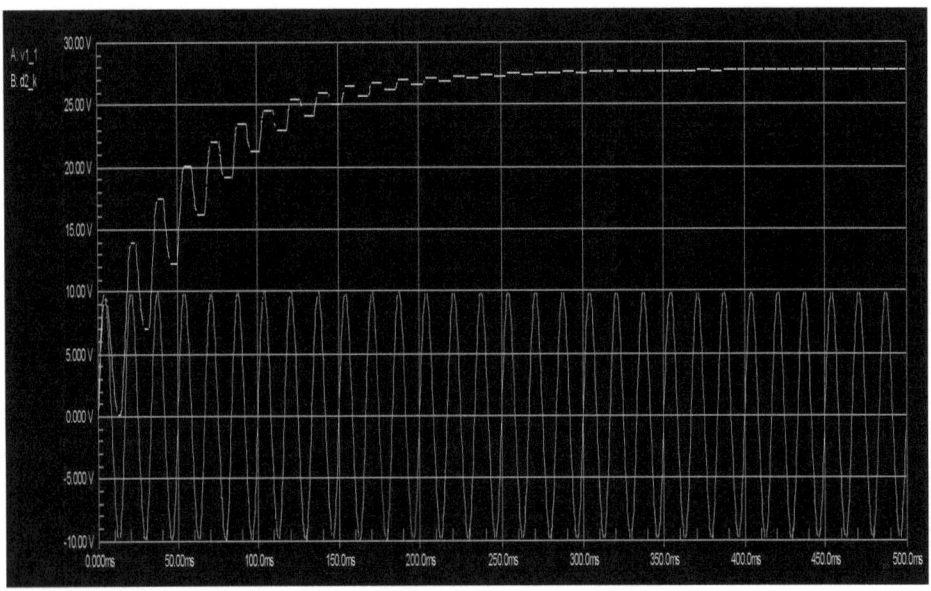

Voltage Tripler

In the circuit on the previous page, a simple half-wave rectifier circuit is followed by a half wave voltage doubler, giving three times the peak voltage less the losses incurred in the diodes and capacitors.

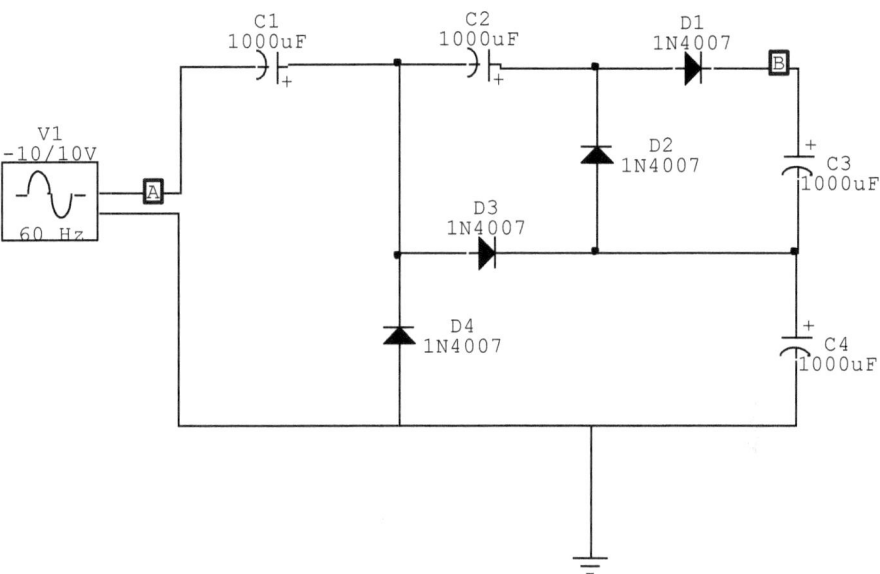

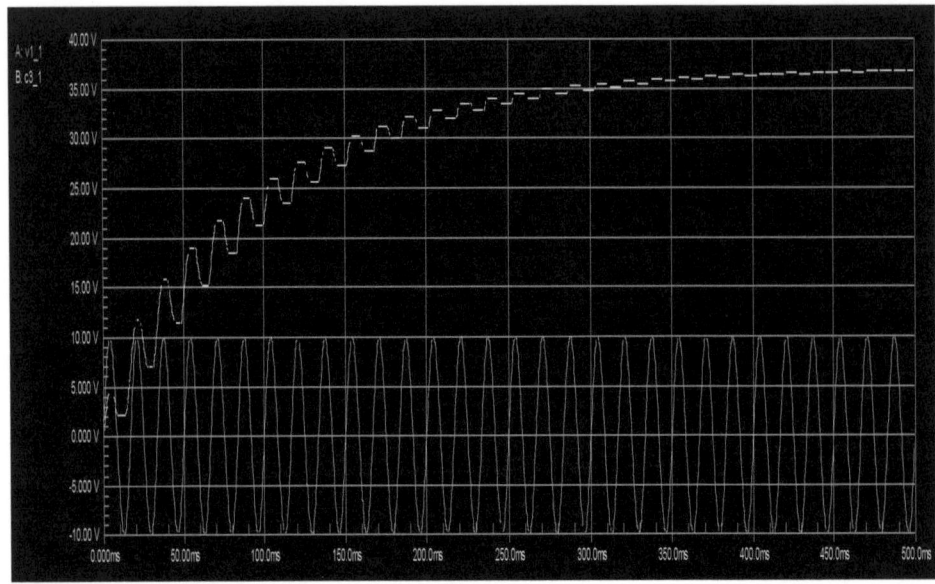

Voltage Quadrupler

The circuit shown on the previous page consists of two full wave doublers added together. The calculations for this type of circuit say an input voltage of 10 volts peak is multiplied by 4 giving 40 volts. In actual fact however, with the losses through the capacitors and the diodes, the actual unloaded output voltage never goes above 35 volts.

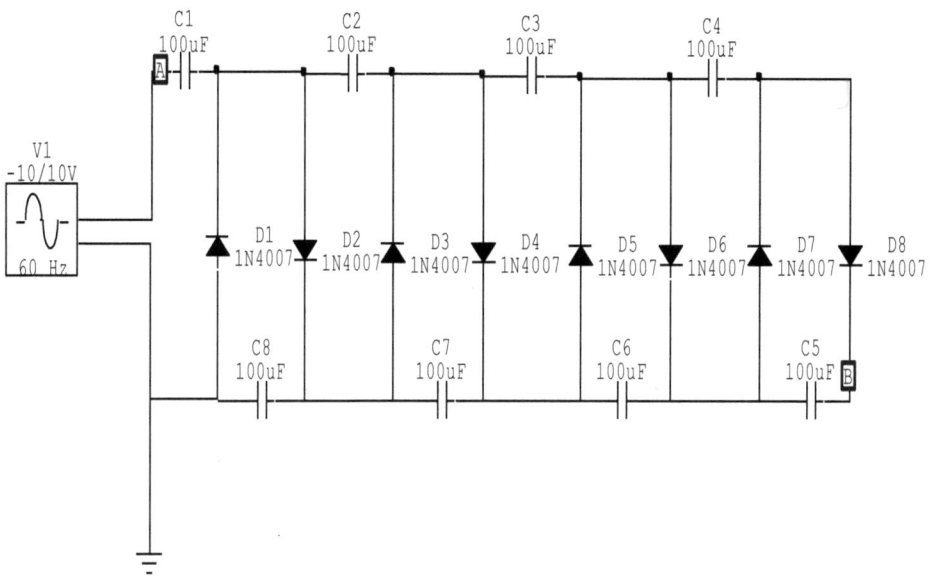

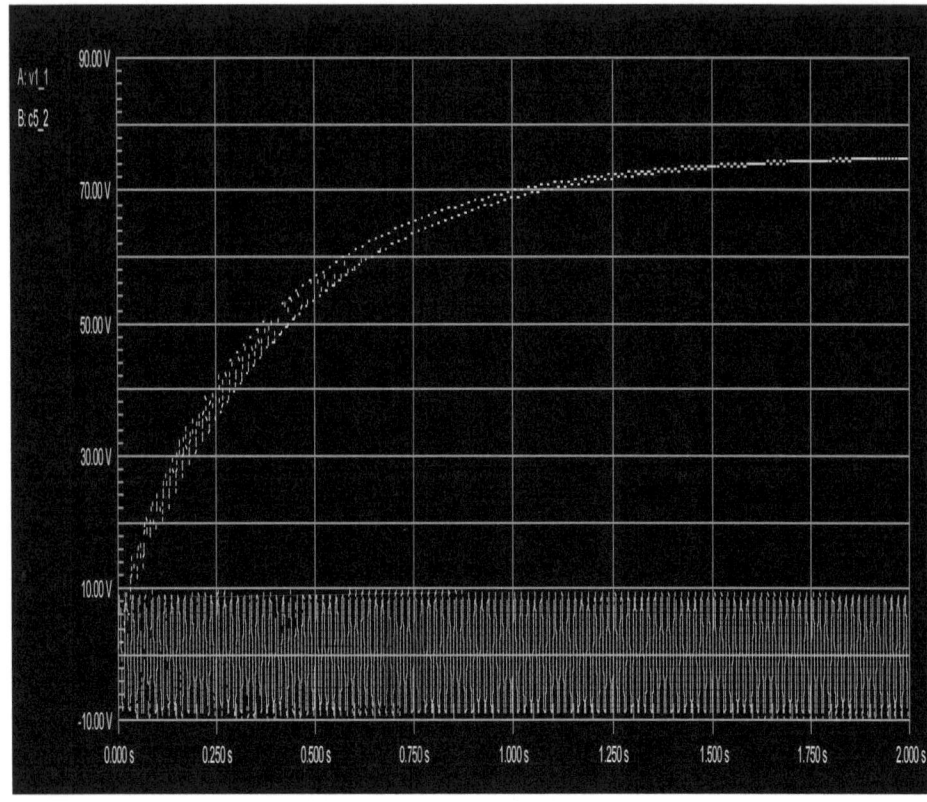

The Cockroft Walton Voltage Multiplier

The Cockroft Walton voltage multiplier shown on the previous page is basically a series of cascaded half wave voltage multipliers. This type of power supply is usually used for high voltage low current applications. They see use in air ionizers, x-ray machines, photocopiers and laser printers. Nobel Prize winning physicists John Cockroft and Ernest Walton used a large version of this device in 1932 as a power supply for their particle accelerator which performed the first nuclear particle disintegration in history. It was named after them even though the device was first invented in 1919 by Swiss physicist Heinrich Greinacher. The circuit can be driven by an AC or pulsating DC waveform.

The circuit shown in the preceding diagram increases the peak voltage by approximately 8 times. With losses, the unloaded voltage ends up being around 75 volts. Since this circuit can only deliver small currents, it should be loaded with a high resistance, at least 1 megohm so that the full 75 volts can be realized.

Full Wave Cockroft-Walton Voltage Multiplier

The full wave version of this multiplier shown on the following page is more efficient than the half wave multiplier. It can deliver more voltage and current because it uses both the negative and positive going peaks of the sine wave to charge up the capacitors.

Note: These Cockroft-Walton voltage multipliers are capable of developing lethal voltages that can kill. The information provided here is for educational purposes only.

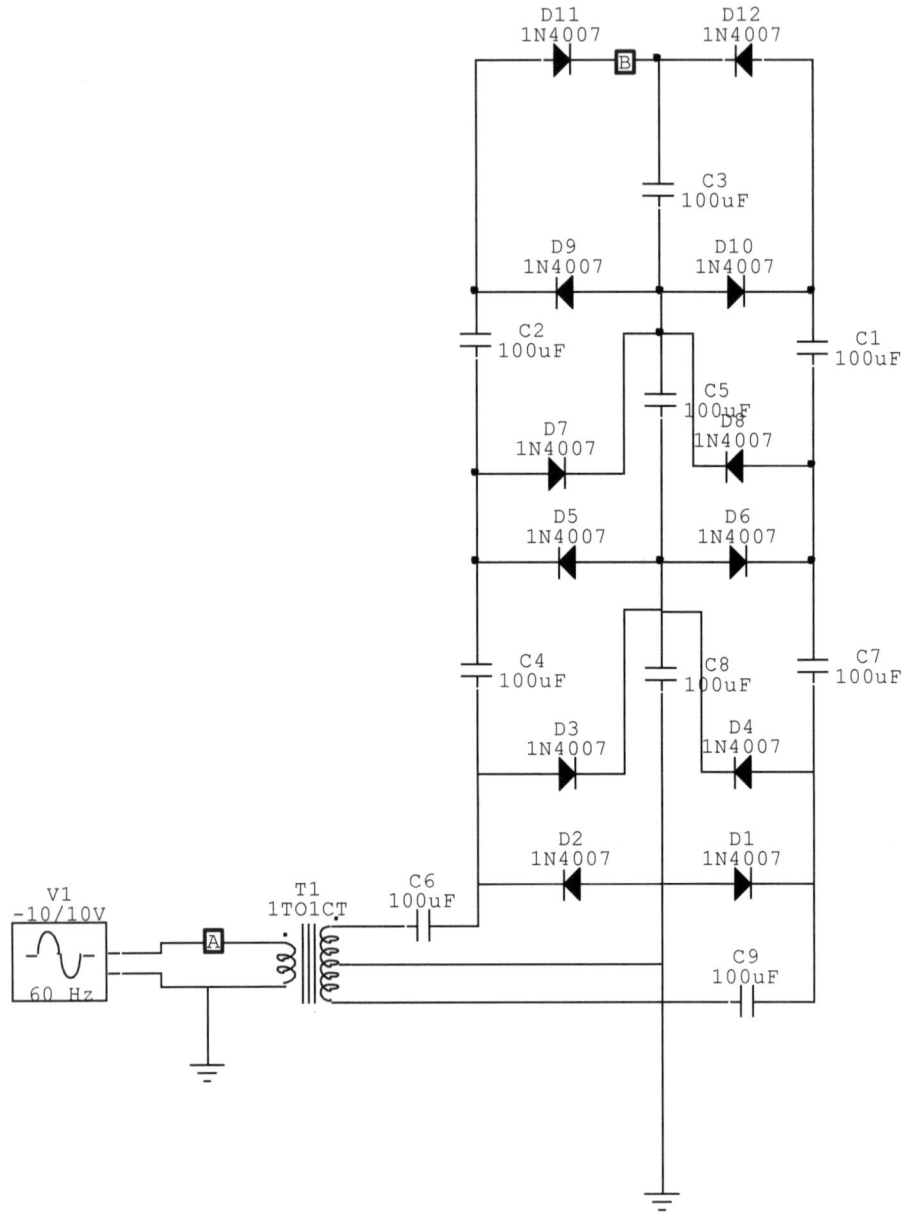

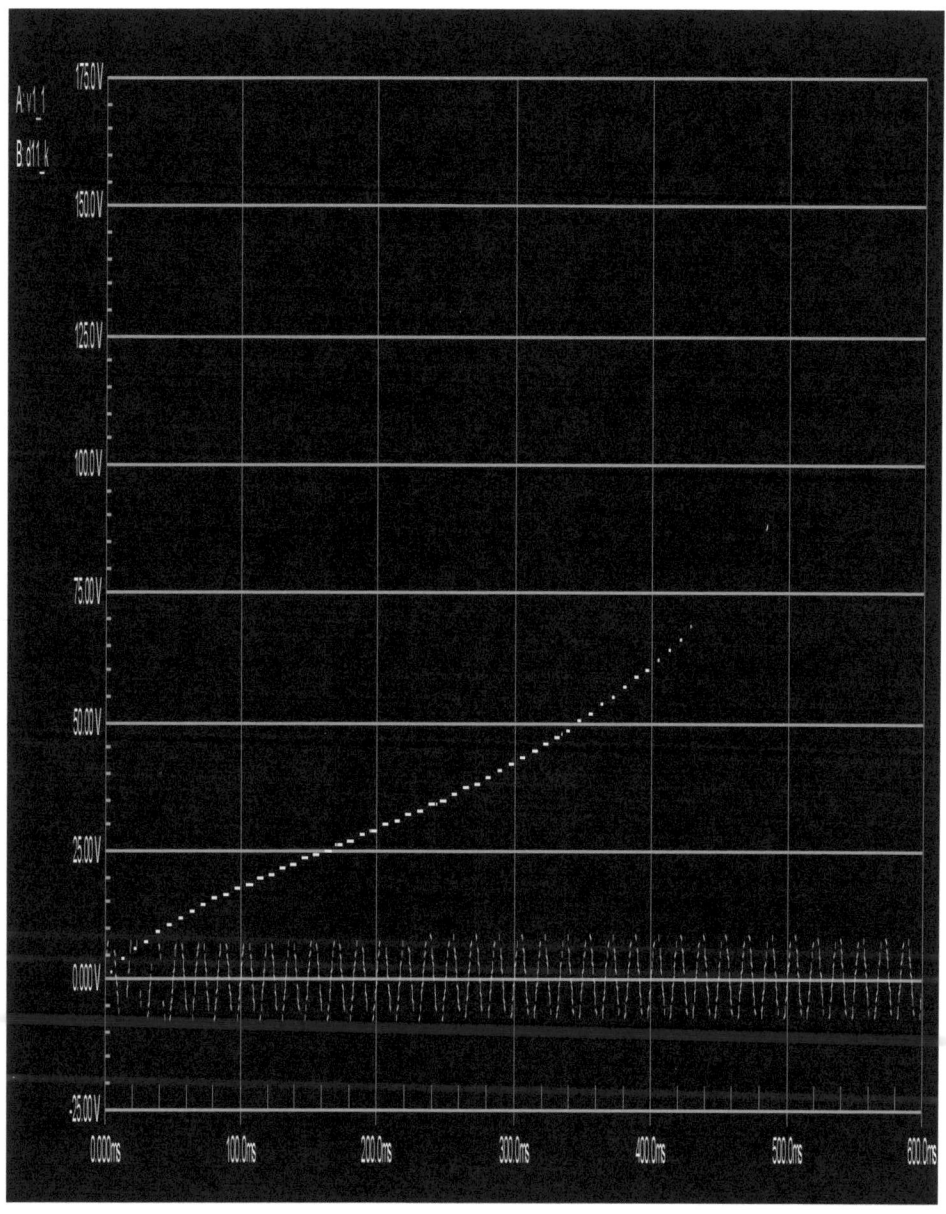

Boost Converter or Charge Pump Circuit

The circuit on the following page takes an AC or positive pulsating DC input that turns on the BS170 Mosfet driving it hard into conduction. The current flows through the inductor from ground to the positive terminal producing a magnetic field. When the Mosfet turns off, the collapsing magnetic field causes a counter electro-magnetic force to flow through the diode and charge up the capacitor. With each succeeding current flow, the capacitor charges up a little bit more positive each time, eventually reaching a higher voltage than the supply. If a higher frequency is used, the inductor used can be smaller, increasing the efficiency.

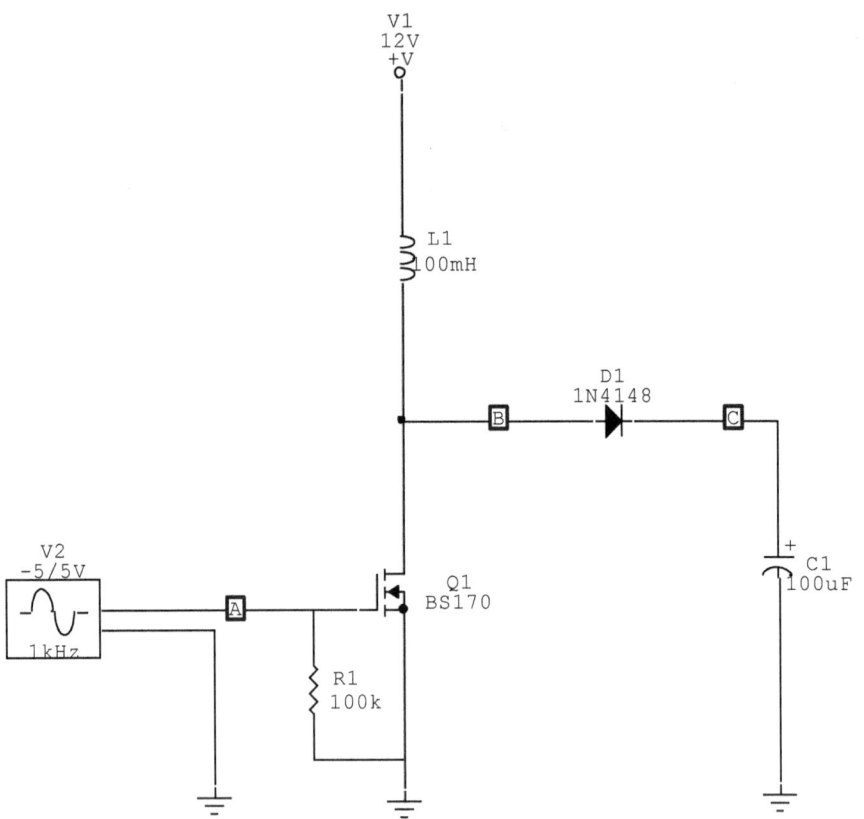

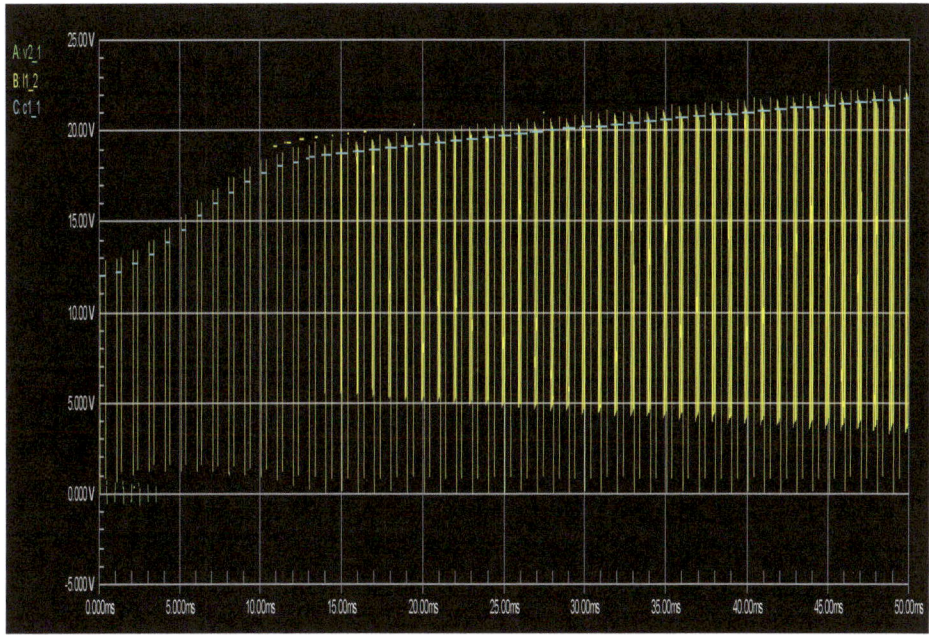

The input signal is green, the waveform at the point where the Mosfet, Inductor and Diode meet is yellow and the blue is the voltage across the capacitor which rides on top of the yellow waveform less the .6 volt drop across the diode. The efficiency of this circuit could be further increased by using a faster switching, lower voltage diode like a Schottky.

Diodes used as circuit protective elements

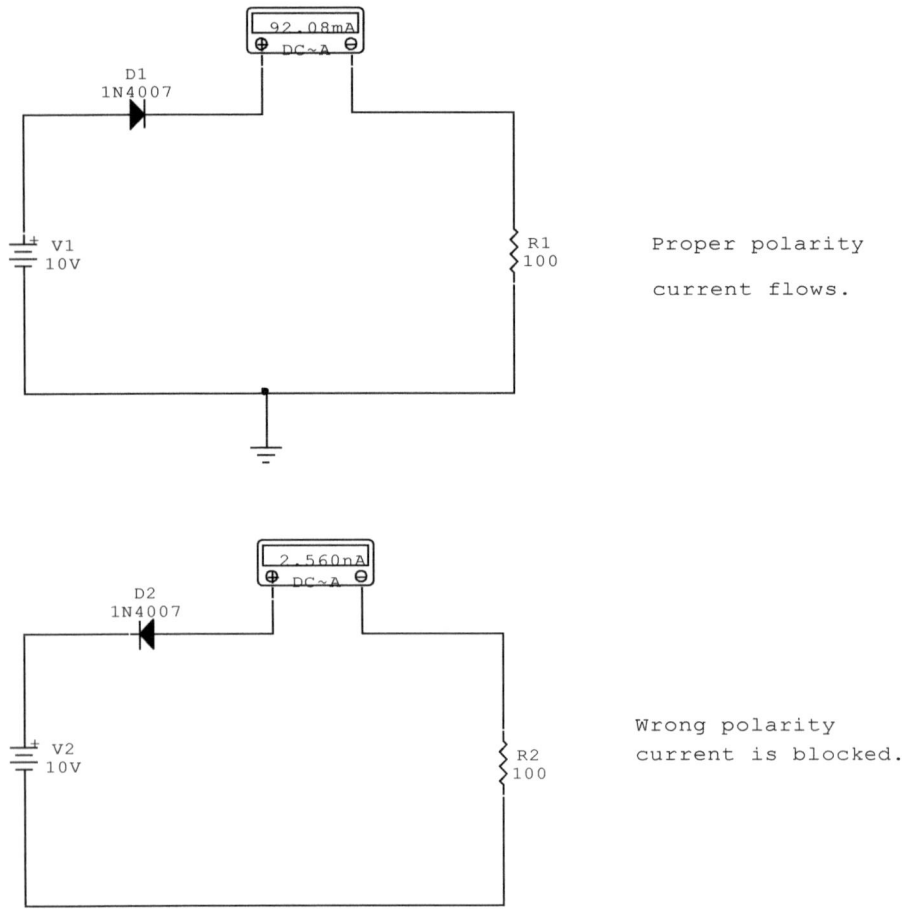

Proper polarity current flows.

Wrong polarity current is blocked.

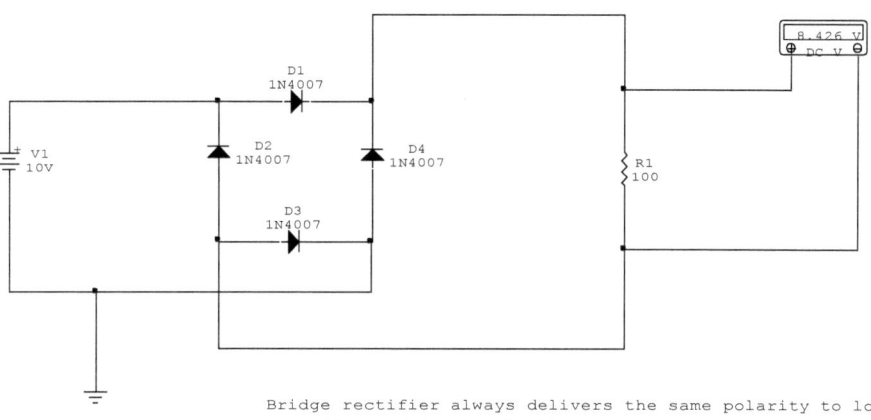

Bridge rectifier always delivers the same polarity to load, regardless of input polarity.

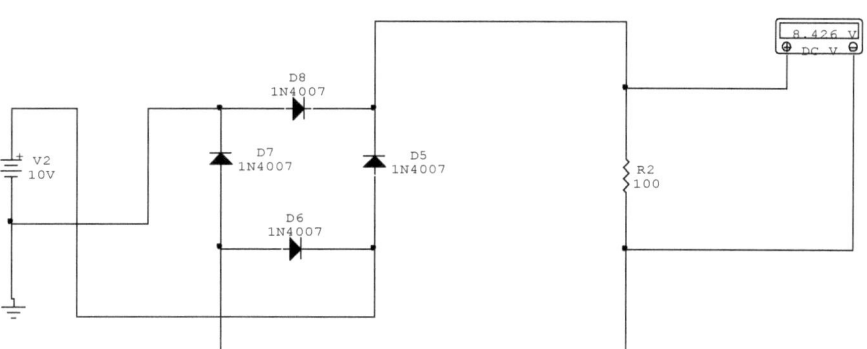

Frequency multiplier using the 1N914/1N4148

The 1N914 switching diode can be employed as an effective frequency multiplier which can be used to extend the range of cheap signal generators. With a 2 volt peak to peak input at 1 MHz, a useful comb of frequencies is generated up into the hundreds of megahertz. Each frequency is separated by 1 Mhz. With a 2 volt peak to peak input signal at 10 Mhz, the comb frequencies are separated by 10 MHz. This circuit is called a "Comb Generator" and is used extensively by industry to do Radio Frequency Interference (RFI) testing to consumer and medical electronics. There are many other uses for this device in areas such as the microwave industry where a more sophisticated circuit would be used to generate much higher frequencies. There are even comb generators used in the fiber optic industry to produce multiple wavelengths of light.

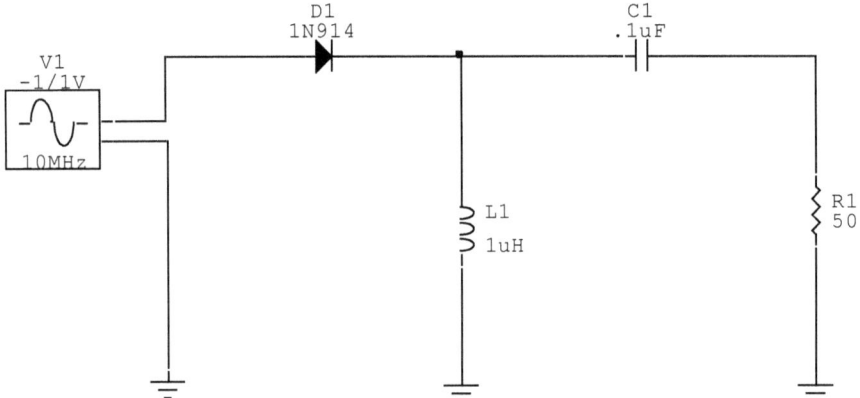

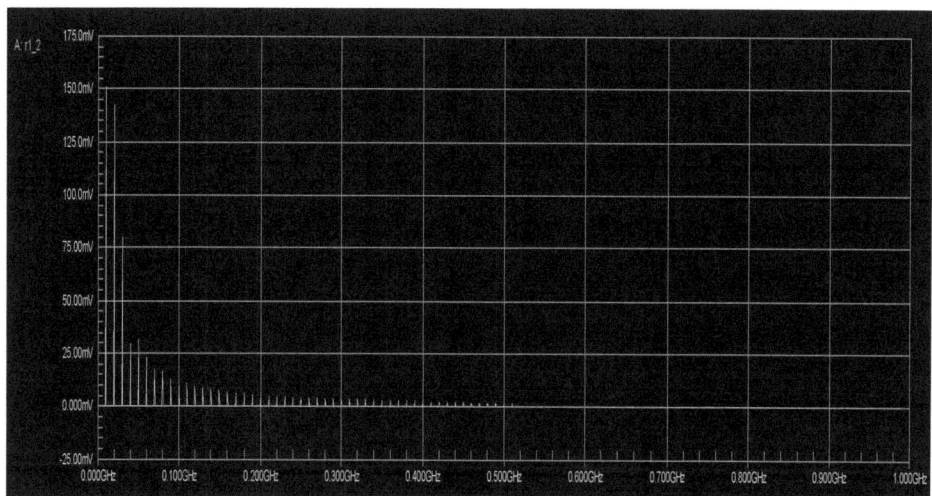

1 GHz frequency sweep of circuit output with 10 Mhz input signal.

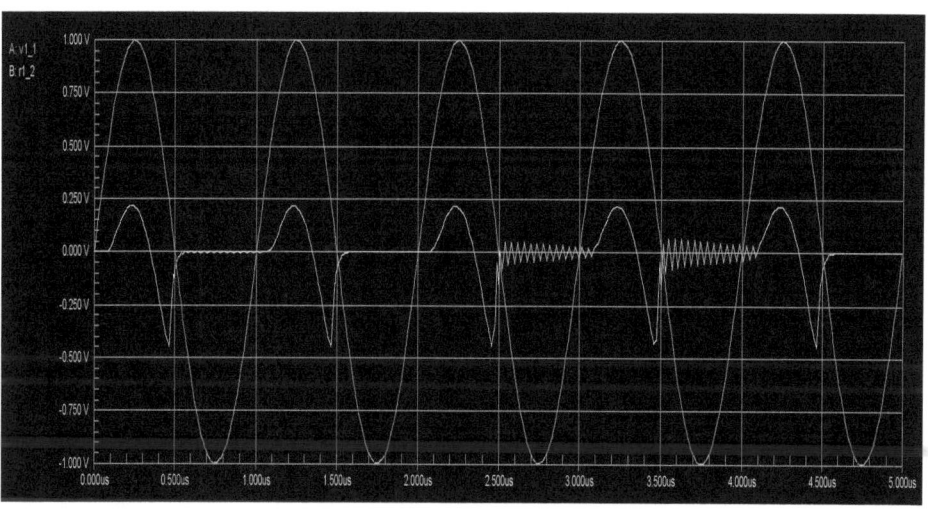

Waveform (green) is transformed into a highly distorted, harmonic rich signal (yellow).

Getting two outputs with one input.

Using diodes to separate a sine wave into positive and negative going pulses

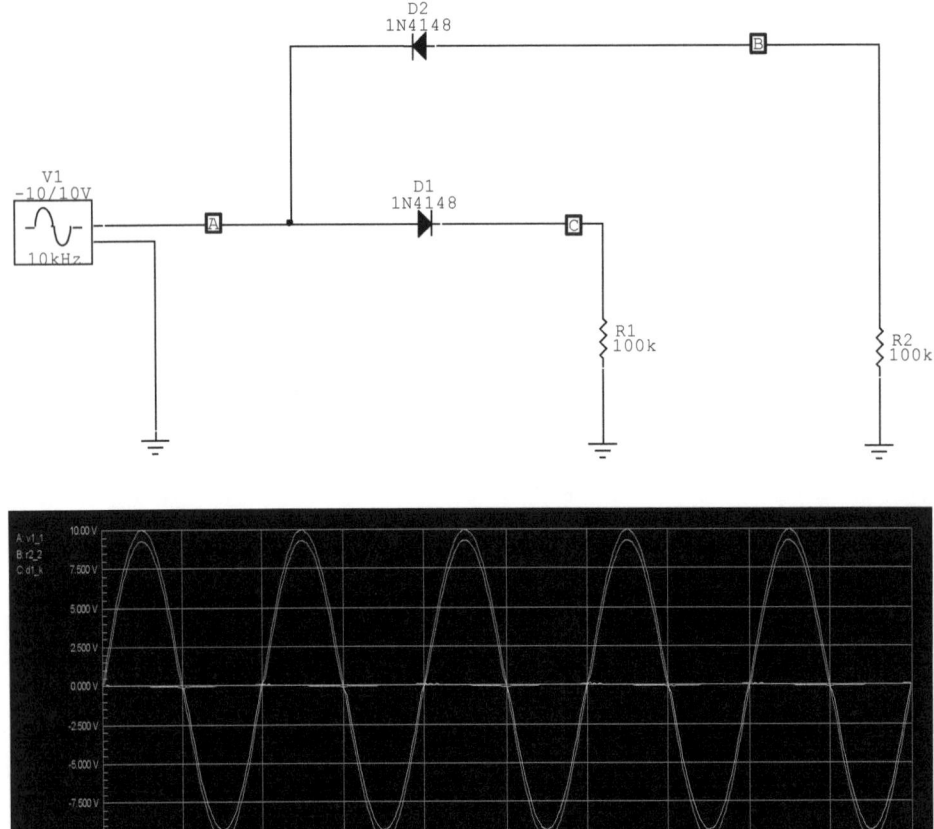

Input sine wave is green and two outputs are blue and yellow.

Reduce or eliminate relay chatter with this circuit.

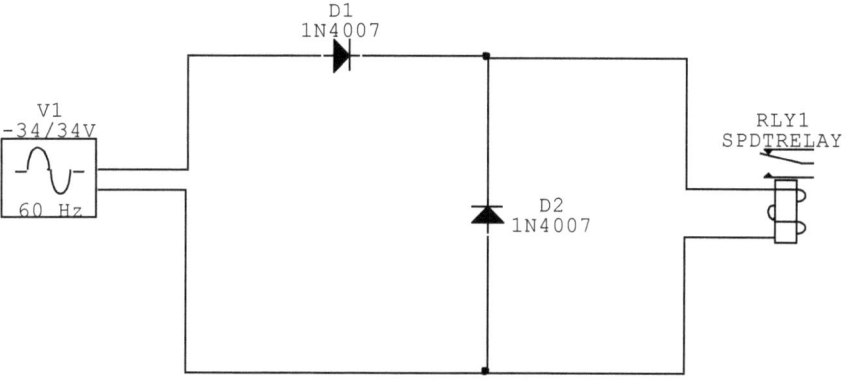

If you have a relay or solenoid that is humming or chattering due to being powered by an AC source, the noise can be reduced or eliminated by adding a couple of diodes in a configuration such as this. The 1N4007 diode has a current rating of 1 amp so if the relay has a greater rating just replace with higher current diodes accordingly. Note that the voltage used in the schematic is the peak voltage for 24 volts RMS, a common voltage rating for relays and solenoids.

Envelope Detector

This circuit is the classic AM detector circuit where the modulation envelope has it's bottom cut off by the diode and the RF component is bypassed by the capacitor leaving only the modulation to appear across the resistor. In this circuit, we are using a 1000 Hertz tone to modulate a 10 kilohertz carrier for illustration purposes. In the first picture, the modulation envelope can be seen in green and the demodulated signal can be seen as yellow. The second picture shows the demodulated signal only.

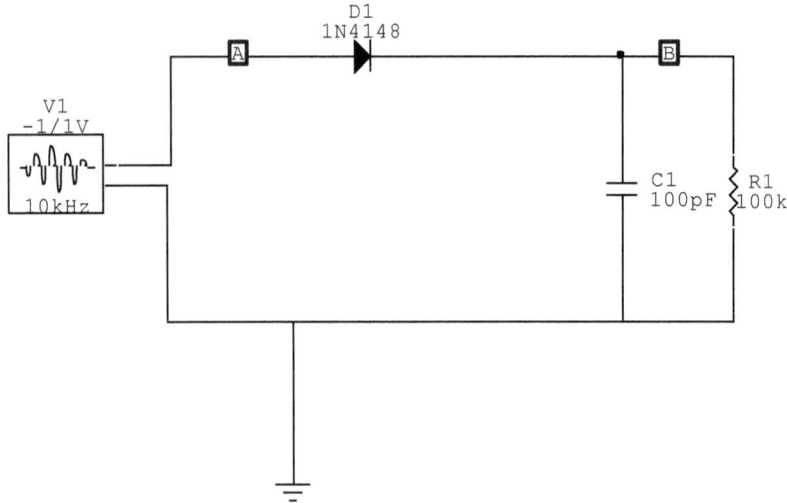

If a tunable input coil, an antenna and ground were added to this circuit, we would have an old fashioned "Crystal Radio" circuit.

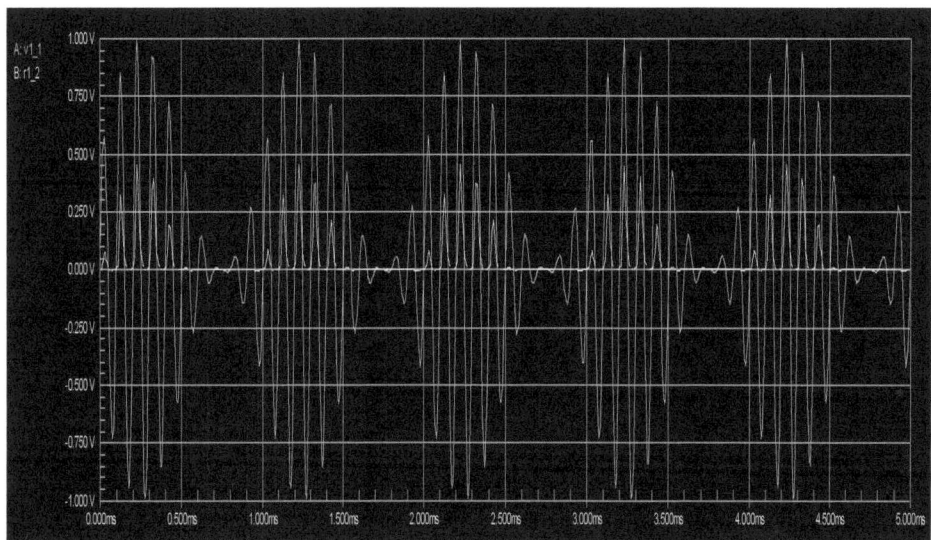

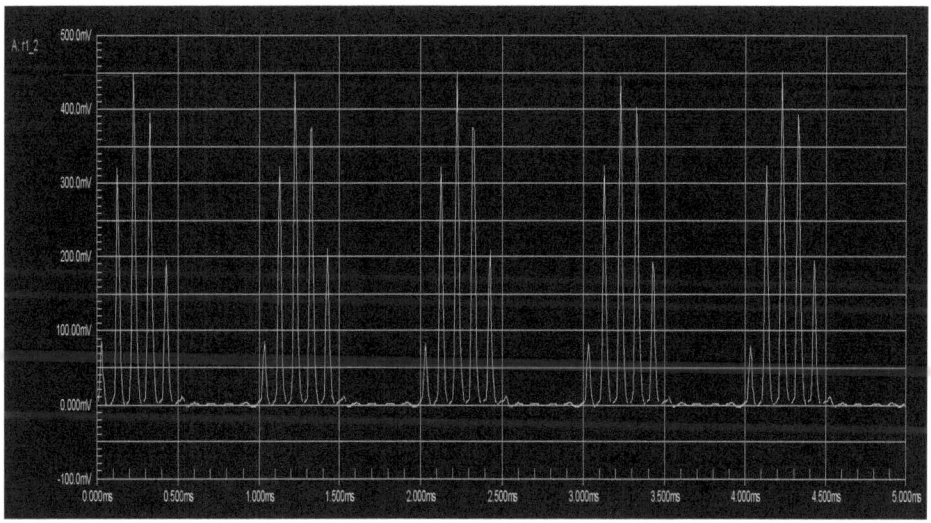

Zener Diode Operation

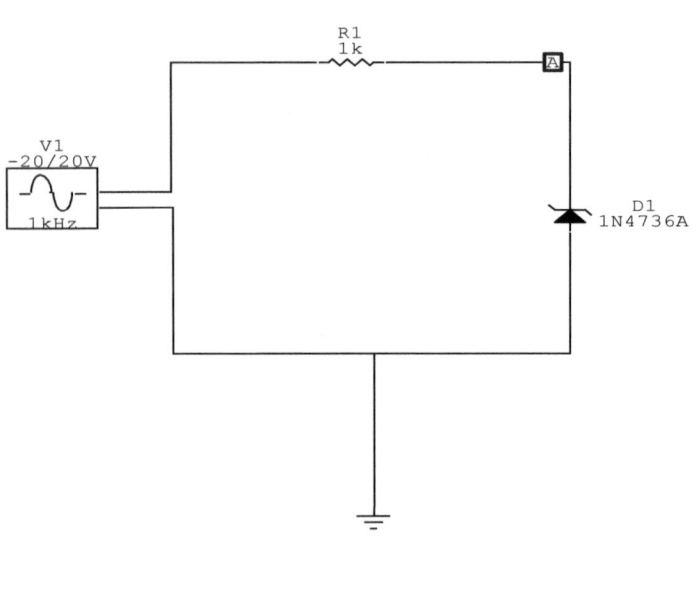

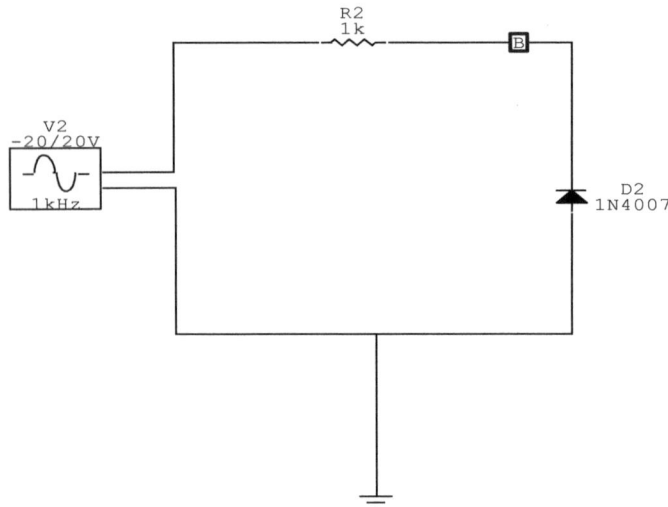

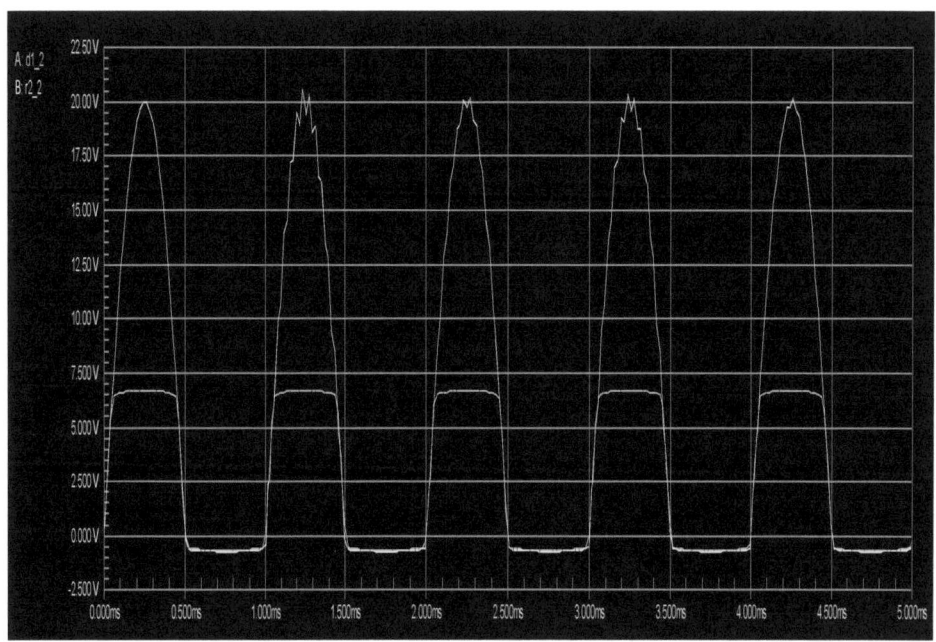

This picture compares the action of a 40 volt peak to peak signal across a 1K resistor and a 6.8V Zener diode (green) and a conventional diode (yellow). The conventional diode (1N4007) cuts off one half of the sine wave giving only a positive going pulse. The Zener diode acts as a regular diode cutting off the lower part of the sine wave but cuts off the upper part of the waveform as it goes past 6.8 volts. The zener is acting as a pretty good square wave generator.

Simple zener diode voltage regulator

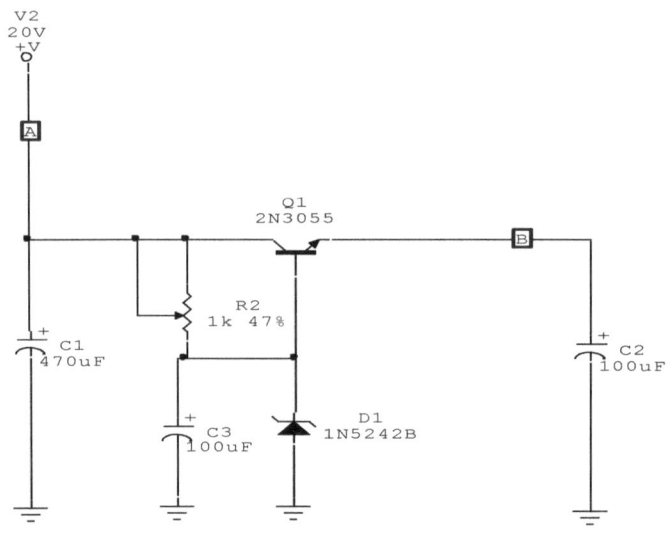

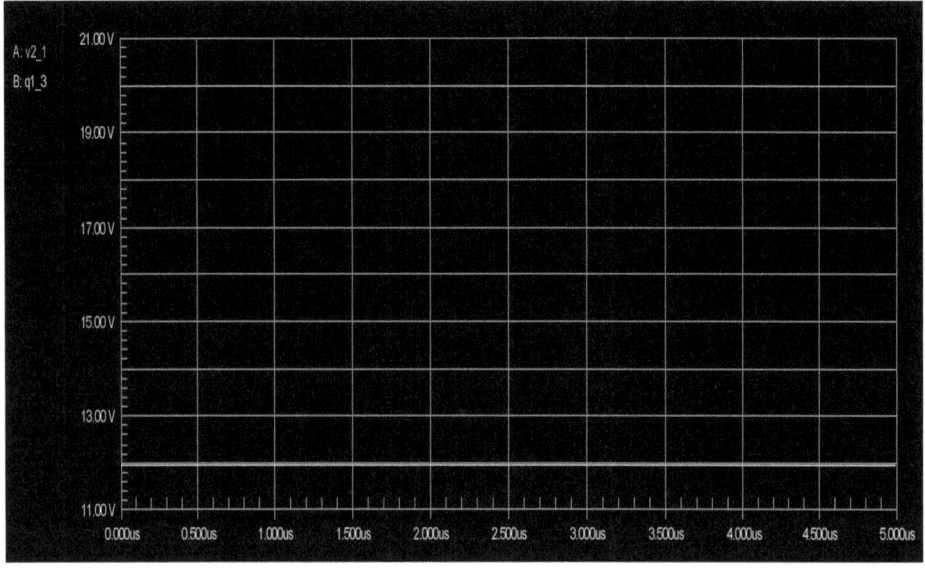

Looking at the schematic on the previous page, the rectified input voltage appears across capacitor C1 and is dropped from 20 volts to 12 volts across the emitter-collector junction of the 2N3055 series pass transistor.

The zener diode D1 sets the voltage at the base of the transistor at 12 volts and the capacitor stabilizes this voltage and bypasses any noise produced by the zener diode. The potentiometer allows fine voltage adjustment depending on the load. The output capacitor stabilizes the output voltage and further bypasses any noise that may have passed through the voltage regular or produced by the regulator itself. This regulator is called a "series pass type regulator" and has a low efficiency compared to a switching power supply. The efficiency of the former is 30-40 percent as opposed to the 90 percent efficiency of the switching power supply. The voltage dropped across the transistor is dissipated as heat, lowering the efficiency. The main advantage to this type of regulator is the much lower noise production (as opposed to the enormous amount of noise produced by a switching power supply) and no need to have the extra expense of the special filtering circuits required in a switching power supply.

Crowbar circuit using a zener diode and Mosfet

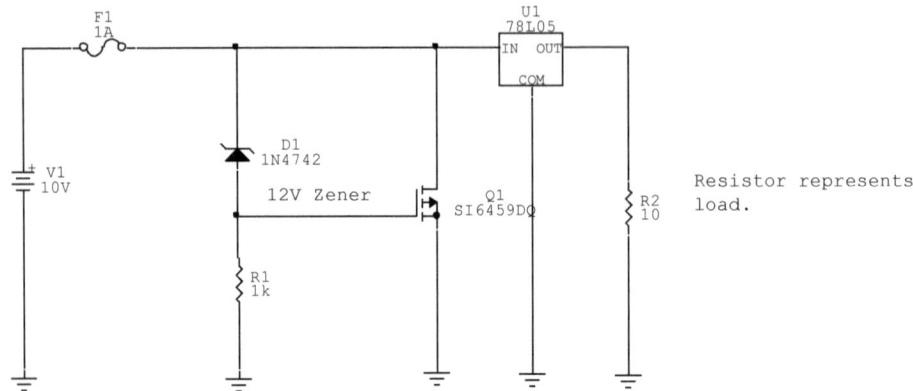

This circuit can be used to provide over voltage protection to circuits. In this configuration, a 12 volt zener diode will fire if the voltage of the power supply goes above 12 volts. When it fires, it turns on the mosfet, creating a short across the input supply blowing the 1 amp fuse. This gives almost instant protection to the voltage regulator. The mosfet should have a current rating that is a number of times higher than the rating of the fuse in order for the circuit to work properly.

Crowbar circuit using a zener and SCR

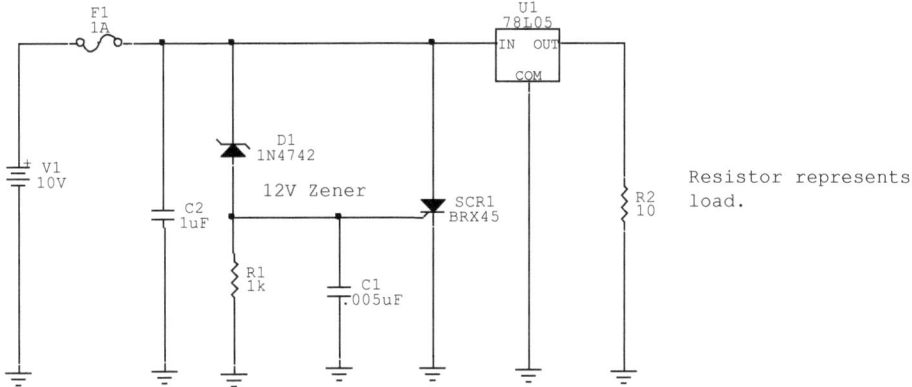

This circuit is very similar to the previous one except it uses a silicon controlled rectifier (SCR) instead of a mosfet. As in the previous circuit, the 12 volt zener diode fires if the input voltage goes above 12 volts turning on the SCR and blowing the fuse. The input capacitor C2 bypasses small insignificant voltage spikes that might trigger the circuit. Capacitor C1 is a "snubber capacitor" that keeps the SCR from triggering on power up. The value of this capacitor is quite critical in that if the value is too high it could delay the operation of the SCR. It should be low enough to still be able to filter the transients that could trigger the SCR. The rating of the SCR should be large enough to blow the fuse. An SCR rating of 3 amps should be more than sufficient to blow a 1 amp fuse.

Tips on using zener diodes

Isolate the zener diode circuit with an emitter or source follower. For best stability of the zener voltage, the current flowing through the zener should be kept constant. Any variations in current consumed by the load must be isolated from the zener so as to keep the zener voltage as constant as possible. By following the zener circuit with an emitter or source follower circuit, the variations that the zener sees will be minimized.

Feed with a constant current source for best stability. Most zener diodes have a simple resistor in series with them but for even greater stability, a constant current source can be used as the supply.

Sufficient current is required to maintain reverse breakdown.
A 1 watt zener diode requires around 12 mA to keep the diode in the reverse breakdown state. The specifications sheet will give information about a particular device.

Maximum current through the zener cannot be exceeded. The specification sheet needs to be consulted so that the maximum current ratings aren't exceeded. As with any semiconductor device, the device can be easily damaged by heat.

Zeners of around 5 volts have the best stability. If voltage stability needs to maintained over variations of temperature best results will be obtained with zener diodes rated around 5 volts.

Amplitude modulation using one diode

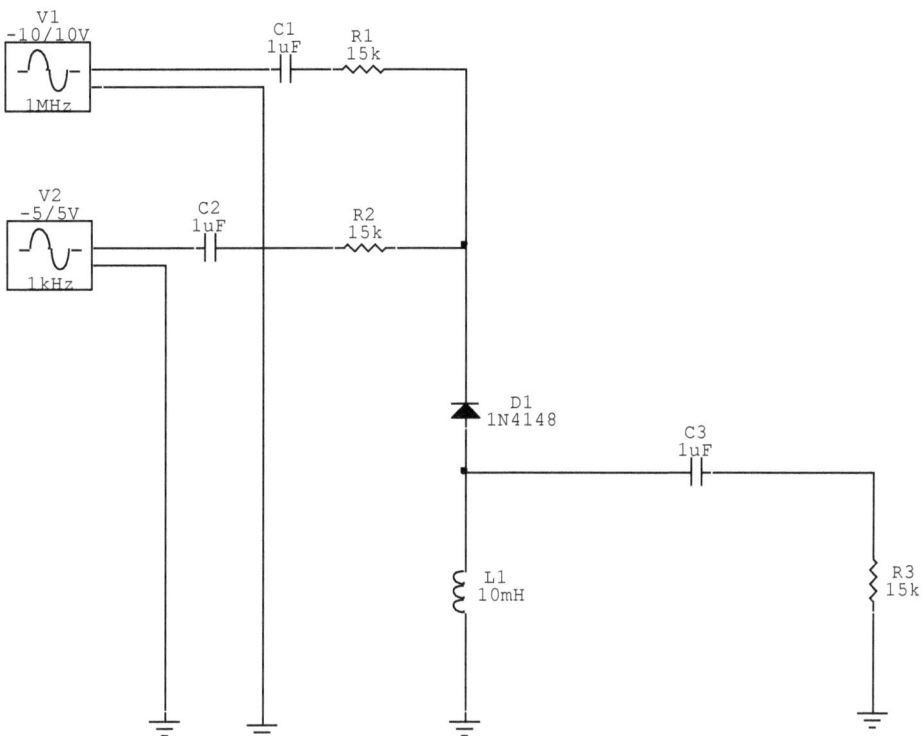

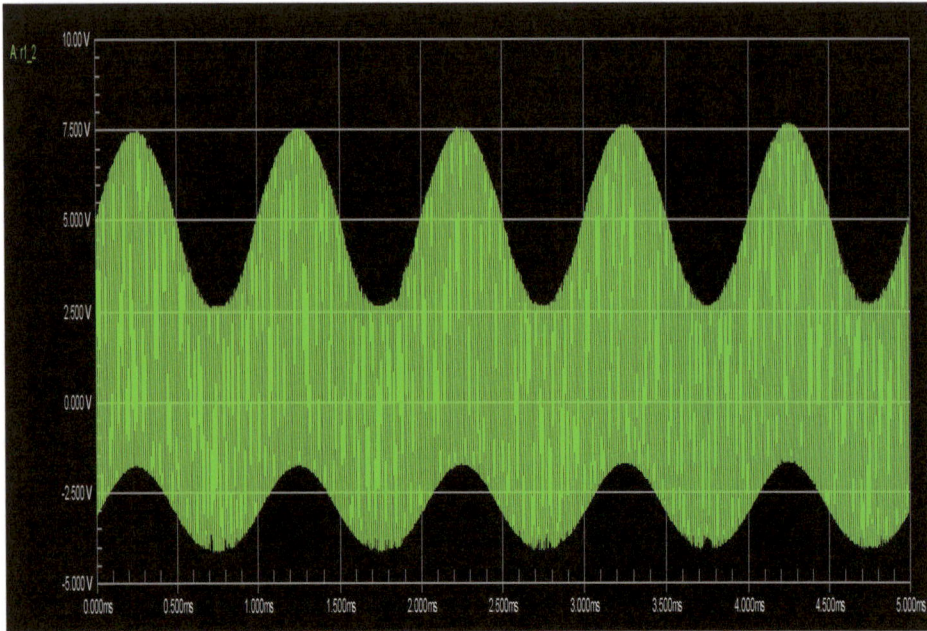

Waveform taken at the point where the two 15k resistors are connected together. The 1kHz waveform can plainly be seen within the 1MHz signal. In addition, there is a DC component in this signal. The diode rectifies these two input waveforms and forms the crudely modulated signal we see in the above graphic.

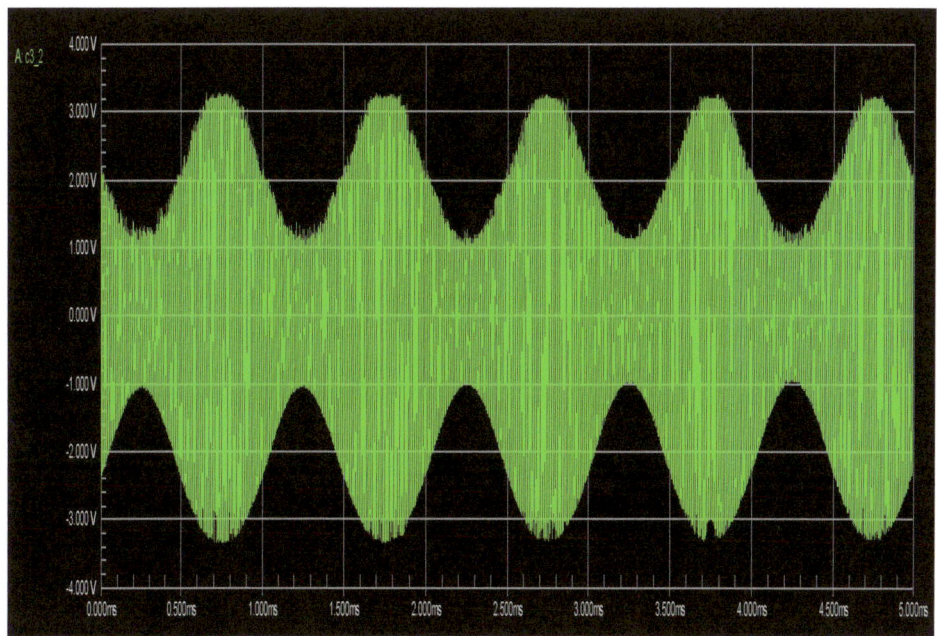

The inductor acts as a low impedance load for the diode and the capacitor passes the signal onto the 15k resistor while blocking the DC component, giving us the properly modulated AM signal we see here. The above signal is modulated at approximately 50% or a .5 modulation index. A 100% modulated signal would result in the upper and lower troughs almost touching each other.

Ring Modulator

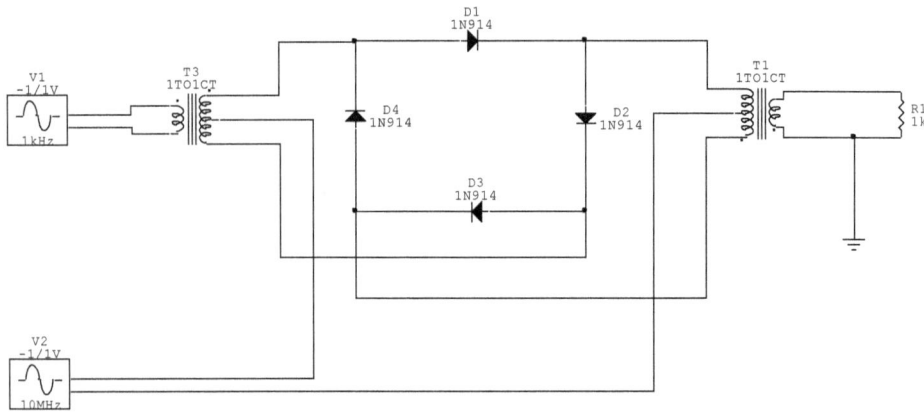

This type of modulator is used to produce a Double Sideband Suppressed Carrier signal. It bears a resemblance to the classic bridge rectifier circuit but is different in the configuration of the diodes and in the use of input and output transformers. In the ring modulator, the diodes are arranged in a ring configuration but in series until the fourth diode's cathode meets the first diode's anode. The modulation is applied to the input transformer and the output is taken from the secondary of the second transformer. The carrier is applied to the center taps of the input and output transformers.

This circuit would be found in single sideband transmitters and the output would be followed by a highly selectable crystal or mechanical filter where either the upper or lower sidebands could be selected.

Practical Ring Modulator for SSB

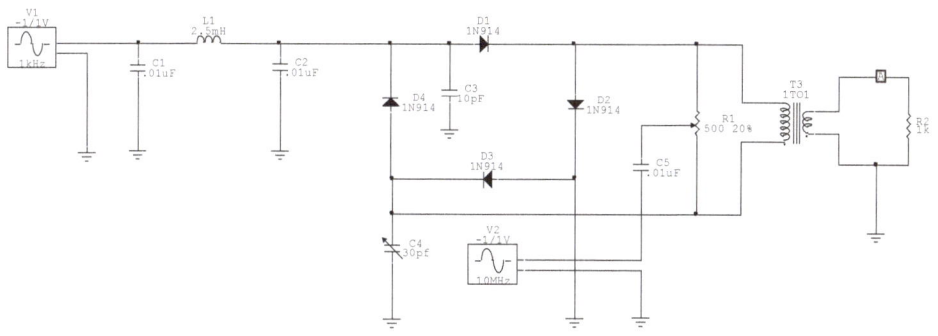

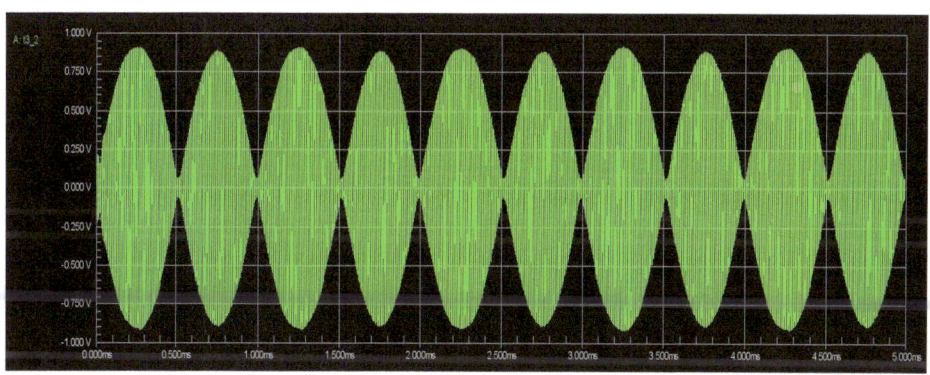

This is a practical working example of a ring modulator which can be incorporated into a transmitter design. R1 is the balance control for the output signal. The output signal shown is set at approximately 100 percent modulation. The output of this circuit would be passed onto a crystal filter where either the upper or lower sidebands could be selected.

www.ingramcontent.com/pod-product-compliance
Lightning Source LLC
Chambersburg PA
CBHW041942240526
45473CB00033B/382